**Safa Rguez**
**Majdi Hammami**
**Ibtissem Hamrouni Sallemi**

# Nanotechnology in the Encapsulation of Essential Oils

**Safa Rguez**
**Majdi Hammami**
**Ibtissem Hamrouni Sallemi**

# Nanotechnology in the Encapsulation of Essential Oils

**ScienciaScripts**

**Imprint**

Any brand names and product names mentioned in this book are subject to trademark, brand or patent protection and are trademarks or registered trademarks of their respective holders. The use of brand names, product names, common names, trade names, product descriptions etc. even without a particular marking in this work is in no way to be construed to mean that such names may be regarded as unrestricted in respect of trademark and brand protection legislation and could thus be used by anyone.

Cover image: www.ingimage.com

This book is a translation from the original published under ISBN 978-620-6-70497-3.

Publisher:
Sciencia Scripts
is a trademark of
Dodo Books Indian Ocean Ltd. and OmniScriptum S.R.L publishing group

120 High Road, East Finchley, London, N2 9ED, United Kingdom
Str. Armeneasca 28/1, office 1, Chisinau MD-2012, Republic of Moldova, Europe
Printed at: see last page
**ISBN: 978-620-7-73125-1**

# Foreword

Welcome to the captivating world of "Nanotechnology in Essential Oil Encapsulation". This book explores the innovative frontiers where nanotechnology and essential oils meet, unveiling a world of fascinating possibilities for science, industry and health.

Essential oils, precious distillations of nature, have long been celebrated for their therapeutic properties, intoxicating aromas and varied applications. Nanotechnology, meanwhile, is emerging as a revolutionary force, offering molecular-scale solutions to diverse challenges.

In this work, we delve into the fundamentals of essential oils, exploring their origins, therapeutic properties and current applications. We highlight their vulnerability and the imperative need to encapsulate them to preserve their integrity.

Nanotechnology becomes our guide as we detail encapsulation techniques, from nanoparticles to sophisticated material choices, revealing the secrets of protecting essential oils at the nanoscale.

The following pages reveal the revolutionary applications of this alliance between nanotechnology and essential oils. From cosmetics to medicine, via the food industry, nanotechnological encapsulation opens doors to significant improvements, from skin penetration to targeted treatments and flavor stabilization.

However, our exploration doesn't stop there. We also confront the ethical and environmental challenges, highlighting the crucial importance of guiding this technology towards sustainable and ethically responsible practices.

The following pages are a call to the future, an invitation to imagine the discoveries to come and the innovations that will shape our world. The future prospects for nanotechnological encapsulation of essential oils reveal infinite potential, fueled by research, innovation and a commitment to a better future.

We invite you to dive into this captivating journey, exploring the intersections between nature and technology, molecules and nanomaterials. Prepare to be amazed by the discoveries to come, and inspired by the far-reaching implications of this exciting convergence.

May this exploration enlighten your understanding, stimulate your curiosity and spark your passion for the extraordinary possibilities offered by nanotechnological encapsulation of essential oils.

Enjoy your reading!

# Table of contents

## 1. Introduction

Essential oils, aromatic gems extracted from plants, and nanotechnology, an emerging field of infinitely small proportions, join forces in a fascinating ballet at the heart of this book. This alliance, which at first seems improbable, opens the doors to a world of revolutionary possibilities. In this introduction, we explore the nature of this marriage between essential oils and nanotechnology, highlighting the aims of this book and underlining the crucial importance of these two fields in various sectors.

Essential oils, volatile compounds extracted from plants, have played an essential role in human life for millennia. Used for therapeutic, cosmetic, culinary and even spiritual purposes, these plant essences have captured the imagination of many civilizations throughout history. Today, essential oils are attracting growing interest thanks to their therapeutic properties, enchanting fragrances and versatile uses.

On the other hand, nanotechnology is emerging as a scientific discipline that explores and manipulates matter at the nanoscale, where material properties can differ considerably from those at the macroscopic scale. These recent advances open up infinite possibilities for the design of revolutionary materials and technologies. By combining these two fields, nanoscale encapsulation of essential oils becomes an exciting undertaking, offering innovative solutions in a variety of sectors.

This book aims to unlock the mysteries and benefits of using essential oils and nanotechnology together, with a particular focus on the encapsulation process. We will explore different nanoscale encapsulation techniques and their impact on the stability, bioavailability and efficacy of essential oils. By presenting case studies and examining applications in fields such as cosmetics, food and medicine, we aim to provide an in-depth understanding of the possibilities offered by this synergy.

The central aim is to make readers aware of recent advances in this multidisciplinary field, while exploring the ethical and environmental implications of using nanotechnology in the essential oil industry. Through an informative and accessible approach, we aim to enlighten not only health and research professionals, but also enthusiasts passionate about the benefits of essential oils.

Essential oils are more than just aromatic agents; they represent a treasure trove of bioactive compounds with antiseptic, anti-inflammatory and relaxing properties. Their use extends from ancestral healing practices to contemporary innovations in the cosmetics, food and pharmaceutical industries. Similarly, nanotechnology, working on a scale invisible to the naked eye, is revolutionizing materials and applications in fields as diverse as electronics, medicine and the environment.

This convergence between essential oils and nanotechnology opens up unexplored horizons. Nanotechnologies enable more effective encapsulation of the active compounds in essential oils, improving their stability and facilitating their controlled release. This synergy promises significant advances in the creation of products that are more sustainable, more effective and better adapted to the specific needs of each industry.

This book aims to be a bridge between two seemingly disparate but intrinsically linked worlds. By exploring the synergies between essential oils and nanotechnology, we hope to inspire fresh and stimulating thinking about the future of these two fields and the potential innovations they can collectively offer. Let us dive into this captivating exploration of nanotechnology in essential oil encapsulation, where science meets aroma.

## 2. Essential oil basics

### 2.1. Definition of Essential Oils

Essential oils, often referred to as "the soul of the plant", are natural extracts obtained from various parts of plants, such as leaves, flowers, stems, roots and bark. These oils are characterized by their complex composition, comprising volatile compounds such as terpenes, ketones, alcohols and esters. It is these compounds that give essential oils their distinctive aromas and therapeutic properties (Ríos, 2016).

Essential oils are extracted by a variety of methods, including steam distillation, cold expression, maceration and $CO_2$ extraction (Aziz et al., 2018). Each production method influences the chemical composition and, consequently, the final properties of the essential oil (Charles & Simon, 1990). These concentrated extracts have been used for millennia for their multiple benefits, ranging from physical healing to mood enhancement (Mamadalieva et al., 2017).

## 2.2. Origin and Extraction

The origins of essential oils date back to antiquity, when ancient civilizations exploited the aromatic and medicinal properties of plants. The Egyptians used frankincense and myrrh in religious rituals (Mansour et al., 2020)while the Greeks and Romans used essential oils for therapeutic massages (Damian & Damian, 1995). Knowledge of essential oil extraction has evolved over time, from rudimentary methods to more sophisticated processes (Chemat & Sawamura, 2010).

Steam distillation, the predominant technique today, has been perfected over the course of Arab history (Djilani & Dicko, 2012). This delicate method involves the separation of volatile plant compounds by steam, followed by their condensation to obtain the essential oil (Figure 1). Cold expression, used mainly for citrus fruits, involves mechanically pressing the zests to release the oils (Aydeniz-Guneser, 2020).

**Figure 1**. An image of the alembic still, invented by Jabir Ibn Hayyan (El Mostain, 2022)

CO extraction$_2$ , a more modern method, uses supercritical carbon dioxide to extract essential oils without altering their chemical composition (De Oliveira et al., 2019).. These varied methods give rise to a diversity of essential oils, each with its own unique characteristics.

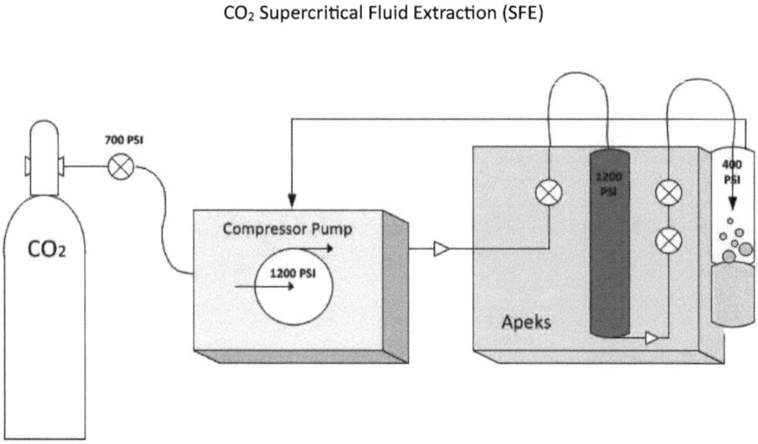

CO$_2$ Supercritical Fluid Extraction (SFE)



**Figure 2**. Schematic diagram of a supercritical CO$_2$ extractor

## 2.3. Therapeutic properties

Essential oils have long been recognized for their therapeutic properties. The compounds present in these oils interact with the body in a variety of ways, offering benefits ranging from stress relief to immune system stimulation (Djilani & Dicko, 2012).

Terpenes, for example, are often responsible for the anti-inflammatory and antioxidant properties of essential oils (Gonzalez-Burgos & Gomez-Serranillos, 2012; F. M. Marques et al., 2019).. Ester-rich essential oils, such as lavender, are prized for their soothing and relaxing properties (Scimeca, 2006). Phenol-based essential oils, such as oregano, have antibacterial and antifungal properties (Laurain-Mattar et al., 2022).

It is essential to note that the therapeutic properties of essential oils vary according to their chemical composition (Lardry & Haberkorn, 2007). Consequently, in-depth knowledge of each oil is crucial for safe and effective use.

## 2.4. Current Applications

Today, the applications of essential oils are diverse and touch on many aspects of daily life. In aromatherapy, essential oils are used to influence emotional well-being, promote relaxation and stimulate energy (Lardry & Haberkorn, 2007). Essential oil diffusers have become a fixture in many homes, creating soothing or invigorating atmospheres.

In the health field, essential oils are integrated into practices such as phytotherapy, naturopathy and traditional medicine (Ali et al., 2015). Their use extends to managing stress, headaches, digestive disorders and even infections (Ali et al., 2015).

In cosmetics, essential oils add a natural and therapeutic dimension to skin care and beauty products (Sarkic & Stappen, 2018). Their antibacterial and antioxidant properties make them valuable ingredients in hair product formulations, lotions and creams (Sarkic & Stappen, 2018).

The food industry is no exception, as essential oils are used to flavor foods, offering natural alternatives to artificial flavors. In addition, their use in food preservation is being explored to extend product shelf life while preserving freshness (Maurya et al., 2021).

In short, essential oils occupy a place of choice in many areas of modern life. Their versatility and therapeutic properties make them invaluable allies, linking a past rich in tradition with a future where nature continues to play a central role. The rest of this journey takes us to the crossroads where nanotechnology meets the very essence of plants.

## 3. Nanotechnology fundamentals
### 3.1. Definition of Nanotechnology

Nanotechnology, a field on the infinitesimal scale, explores and manipulates matter at nanometric dimensions, i.e. on the order of a few nanometers to several hundred nanometers (Silva, 2004). One nanometre is equivalent to one billionth of a metre. At this scale, the physical and chemical properties of materials can differ significantly from those observed on a larger scale

(Zerrougui & Amira, 2023). Nanotechnology encompasses the design, manipulation and use of nanoscale structures and devices to create new materials and develop innovative applications (James, 2023).

## 3.2. Basic principles

The fundamental principles of nanotechnology are based on understanding and manipulating the properties of materials at the nanoscale. Quantum phenomena, such as quantum size and energy quantization, become dominant at this scale. Surface forces, electronic conductivity and magnetic properties may also differ from those of macroscopic materials (James, 2023).

Nanotechnology takes advantage of these principles to create materials and devices with enhanced performance. The ability to control structure at the atomic scale enables the development of new materials with unique properties, paving the way for significant advances in a variety of fields (Dupuy & Roure, 2004).

## 3.3. Applications in Various Industries

Nanotechnology has widespread applications in various industrial sectors, transforming the way we design and use materials (Khandve, 2014). In the electronics industry, the manufacture of nanometric components has enabled the development of smaller, faster and more energy-efficient devices (Pandey, 2022). Nanomaterials are also used in the coatings industry, offering scratch- and stain-resistant surfaces (Khanna, 2008).

In the medical field, nanotechnology is used to design controlled-release drugs, enabling targeted delivery to the body (Venugopal et al., 2008). Nanoparticles can be functionalized to specifically target cancer cells, improving treatment efficacy while reducing side effects (Sinha et al., 2006).

The energy industry is also benefiting from nanotechnology, with applications in the development of materials for more efficient solar cells, more efficient energy storage batteries and improved catalysts for power generation (Deng et al., 2016).

## 3.4. Recent advances

Recent advances in nanotechnology have considerably broadened its field of application. Advanced manufacturing techniques, such as electron-beam lithography and immersion nanolithography, enable greater precision in the creation of nanometric structures (Torres, 2003). Nanomaterials, such as carbon nanotubes and metal nanoparticles, open up new possibilities in the development of lighter, stronger and more conductive materials (Shoukat & Khan, 2021).

Nanotechnologies applied to medicine have seen significant advances with the development of nanorobots capable of specifically targeting diseased cells (Cavalcanti & Freitas, 2005). The fields of regenerative medicine and gene therapy are also benefiting from advances in nanotechnology (Demirer et al., 2021)..

In short, nanotechnology is at the forefront of scientific and technological innovation. Continuous progress in this field promises revolutionary changes in a variety of industries, from medicine to electronics to energy. It is against this dynamic backdrop that the encounter between nanotechnology and essential oils opens up exciting prospects and innovative possibilities. The remainder of this book will explore the fusion of these two worlds at the nanoscale, laying the foundations for a new era in the encapsulation of essential oils.

## 4. Essential Oil Encapsulation Requirements
### 4.1. Vulnerability of Essential Oils

Despite their many beneficial properties, essential oils have a certain vulnerability which limits their use in certain applications. These volatile, sensitive compounds are susceptible to factors such as light, oxygen, heat and humidity (Jugreet et al., 2020).. These environmental conditions can alter the chemical composition of essential oils, reducing their efficacy and quality.

Oxidation is one of the main challenges facing essential oils. When exposed to air, they can undergo oxidation reactions that alter their properties and generate undesirable compounds (Turek & Stintzing, 2013). In addition, the volatility of essential oils can lead to a rapid loss of their characteristic aromas when exposed to ambient air.

## 4.2. Encapsulation benefits

Encapsulation offers a promising solution to overcome the challenges associated with the vulnerability of essential oils (El Asbahani et al., 2015). This process involves trapping essential oil molecules in encapsulated structures, usually on a nanometric scale. These capsules can be composed of different materials, such as lipids, polymers or proteins (Majeed et al., 2015).

One of the main advantages of encapsulation is the protection it offers against harmful environmental factors. Capsules act as a physical barrier, preventing direct exposure of essential oils to oxygen, light and other external agents. This prolongs the stability and shelf life of essential oils, preserving their properties intact (Majeed et al., 2015).

## 4.3. Improved Stability and Service Life

Encapsulation of essential oils brings a significant improvement to their stability and shelf life (Pandit et al., 2016). Capsules provide protection against oxidation by preventing direct contact with air. This minimizes the degradation of volatile compounds and maintains the quality of essential oils over an extended period of time (Barradas & de Holanda e Silva, 2021).

What's more, encapsulation offers precise control over the release of the active compounds contained in essential oils. This control enables a gradual and controlled release, optimizing their efficacy in various applications (Pandit et al., 2016). For example, in the cosmetics industry, the encapsulation of essential oils enables a regulated release for optimal absorption by the skin, enhancing therapeutic and aromatic benefits (Silva-Flores et al., 2023).

In addition, encapsulation offers advantages in terms of handling and transporting essential oils. Capsules can be integrated into a variety of formulations, such as creams, lotions, capsules or edible films, facilitating their use in various end products (Guzmán & Lucia, 2021).

In conclusion, the need for encapsulation of essential oils arises from their inherent vulnerability to environmental conditions. Encapsulation offers an innovative solution by protecting these precious compounds, improving their stability and extending their shelf life. This technique opens the way to new application possibilities in a variety of sectors, from cosmetics to food, redefining the way in which we reap the benefits of essential oils. The rest of this book will explore in detail the different nanoscale encapsulation techniques and their impact on the efficacy of essential oils.

## 5. Nanotechnology Encapsulation Techniques

### 5.1. Nanoparticles and Nanocapsules

Nanotechnological encapsulation of essential oils is based on two key concepts: nanoparticles and nanocapsules. Nanoparticles are nanoscale solid structures, while nanocapsules are hollow structures containing active materials, such as essential oils. Both approaches offer distinct advantages in terms of controlled release and protection of active compounds (Suffredini et al., 2013).

Solid nanoparticles can be composed of various materials, such as polymers, lipids or proteins. These materials form a solid matrix into which the essential oil molecules can be incorporated. Nanocapsules, on the other hand, often consist of an outer shell surrounding a liquid core containing the essential oils. This structure offers additional protection, reducing direct interaction with the external environment.

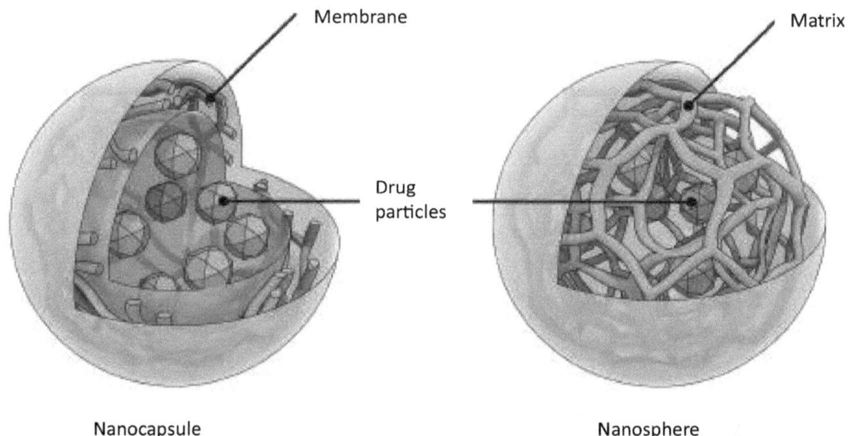


**Figure 3.** Schematic representation of the structure of a nanocapsule and nanosphere. (Suffredini et al., 2013)

## 5.2. Manufacturing techniques

### 5.2.1. Lipid nanoencapsulation

Lipid nanoencapsulation involves the use of lipids as encapsulation vectors. These lipids can be of natural origin, such as phospholipids extracted from lecithin or triglycerides, or synthetic, such as lipid nanoparticles (Assadpour & Mahdi Jafari, 2019). These systems offer excellent solubility for essential oils, protecting their delicate molecular structure. Lipid nanoemulsions, for example, form nanometric droplets stabilized by emulsifying agents, offering improved bioavailability during administration.

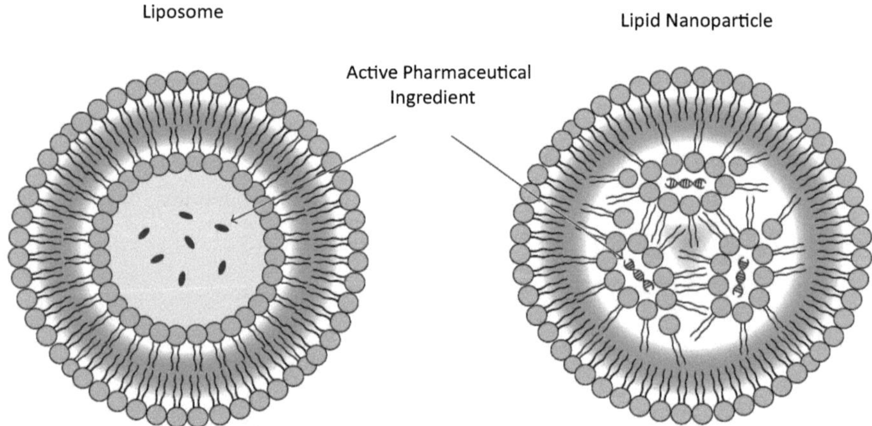


**Figure 4.** Schematic representation of lipid nanoencapsulation

### 5.2.2. Polymeric nanoencapsulation

Biocompatible polymers, such as polyethylene glycol (PEG) or poly lactic-co-glycolic acid (PLGA) nanoparticles, are used to encapsulate essentialoils (Perinelli et al., 2019).. These polymers form stable particles that protect essential oils from external influences and offer precise release control. This method is particularly useful in medical applications, where prolonged, targeted release can be crucial.

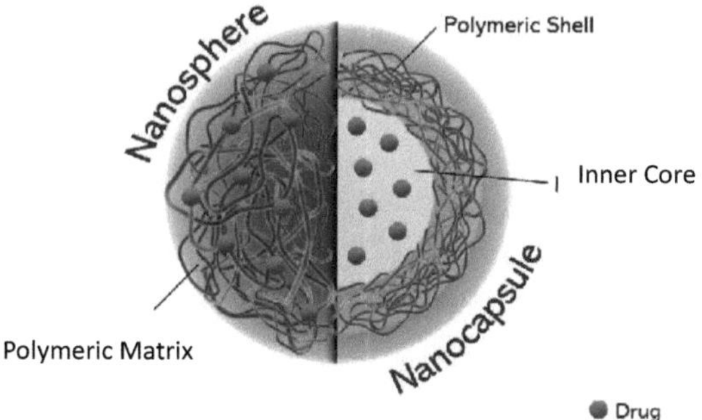


**Figure 5.** Schematic representation of polymeric nanoencapsulation (Gagliardi et al., 2021)

### 5.2.3. Nanoemulsion

Nanoemulsion is a technique that creates stable colloidal systems made up of small droplets of essential oils dispersed in an aqueous phase. The emulsifiers used enable these droplets to be maintained at a nanometricsize (Barradas & de Holanda e Silva, 2021). Nanoemulsification improves the solubility of essential oils in water, making them more versatile for use in a variety of formulations, from personal care products to medicines (Mustafa & Hussein, 2020).

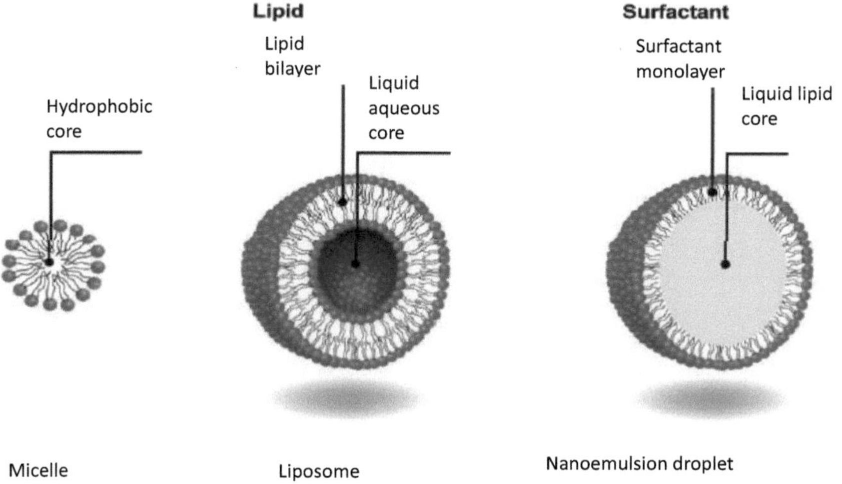


17

**Figure 6.** Schematic representation of the nanoemulsion (Mustafa & Hussein, 2020)

### 5.2.4. Cyclodextrins

Cyclodextrins are toroidal oligosaccharides that form inclusion complexes with essentialoils (Siva et al., 2020). These complexes offer effective protection against oxidation and help mask unwanted odors while improving water solubility. Nanocapsules formed by cyclodextrins are often used in the food and pharmaceutical industries to improve the stability and controlled release of active ingredients (Yuan et al., 2019).

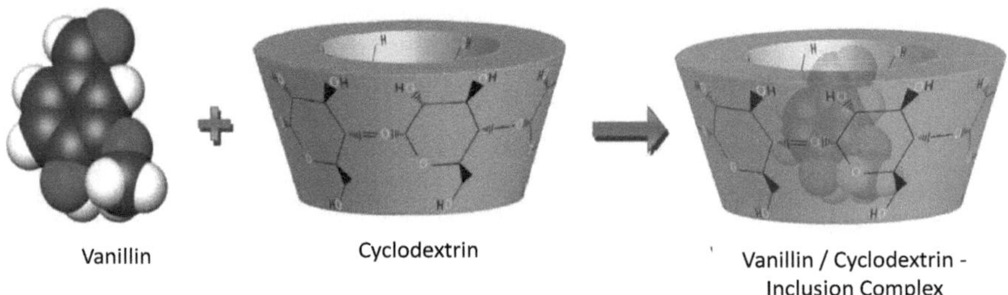


**Figure 7.** Schematic representation of cyclodextrin encapsulation (Kayaci & Uyar, 2011)

### 5.2.5. Nanoencapsulation by complex coacervation

Complex coacervation is a technique where opposing polymers separate and form a shell around the substance to be encapsulated, creating microcapsules or nanoparticles (Timilsena et al., 2019). In the case of essential oils, this method creates a protective envelope around volatile molecules, offering protection against oxidation, degradation and environmental influences.

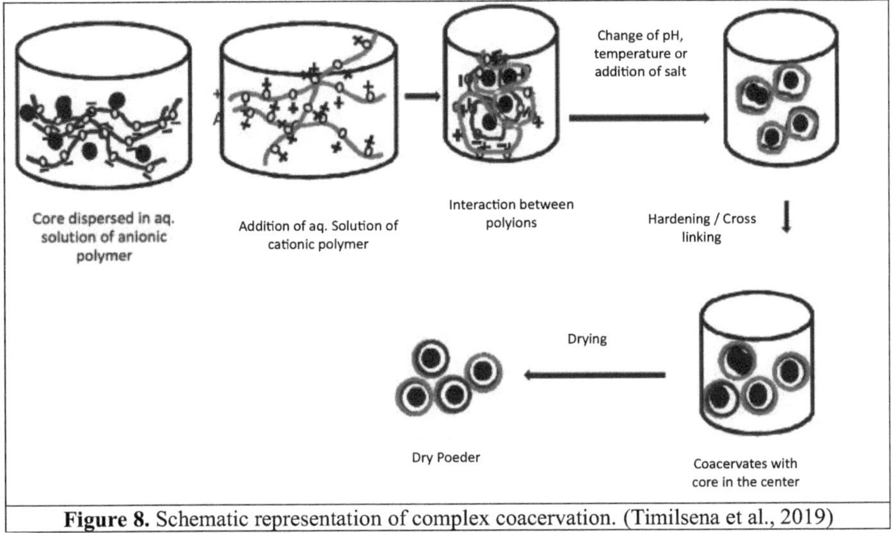


**Figure 8.** Schematic representation of complex coacervation. (Timilsena et al., 2019)

### 5.2.6. Nanoencapsulation by electrospinning

Electrospinning is a method in which a liquid polymer is electrostatically stretched to form ultrafine fibers (Wen et al., 2017). Essential oils can be encapsulated in these fibers to create nanofiber structures. This approach enables controlled release of essential oils and offers a high specific surface area, which can be advantageous in applications such as medical devices or dressings (Wen et al., 2017).

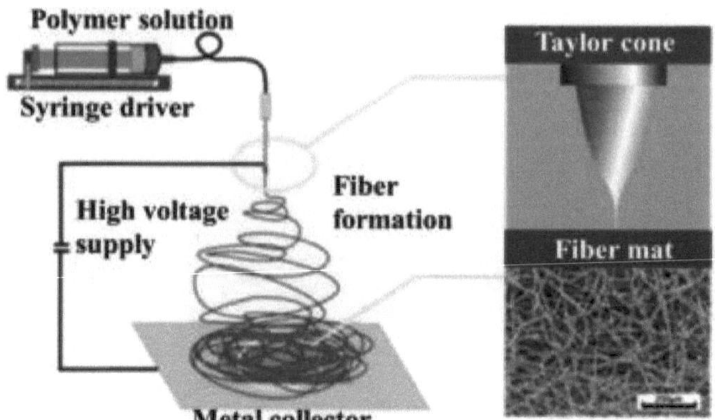



**Figure 9.** Schematic representation of nanoencapsulation by electrospinning

### 5.2.7.  Nanoencapsulation by extrusion

Extrusion is a method where molten or liquid materials are pushed through a die to form continuous structures (Fangmeier et al., 2019). In the context of essential oil nanoencapsulation, this technique can be used to create nanocapsules using thermoplastic polymers. Essential oils are encapsulated in these matrices, offering protection against oxidation and controlled release.

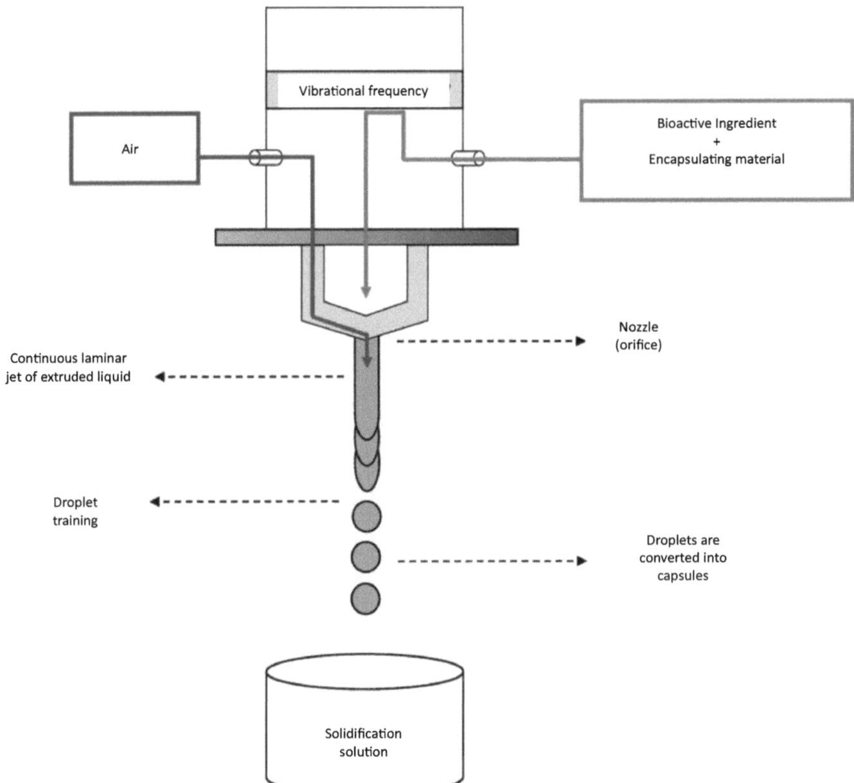


**Figure 10.** Schematic representation of nanoencapsulation by extrusion. (Fangmeier et al., 2019)

### 5.2.8. Magnetic nanoencapsulation

Magnetic nanomaterials, such as iron nanoparticles, can be used to encapsulate essential oils. These magnetic nanoparticles enable external control of essential oil release using a magnetic field (Miguel et al., 2020). This approach can be particularly useful in medical applications to specifically target certain areas of the body.

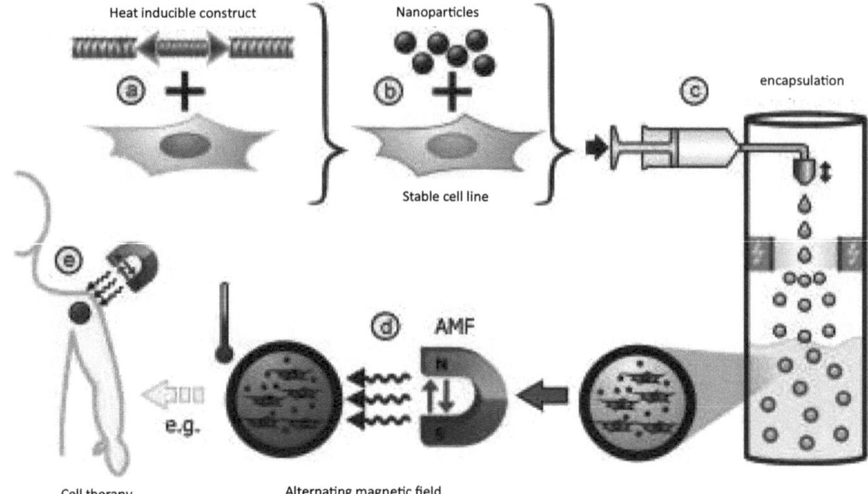


**Figure 11.** Magnetic field-controlled gene expression in encapsulated cells. (Ortner et al., 2012)

By integrating these diverse nanoencapsulation methods, research continues to explore innovative solutions for improving the stability, bioavailability and efficacy of essential oils in a variety of applications ranging from medicine and cosmetics to food and pharmaceuticals. These approaches broaden the range of possibilities for fully exploiting the benefits of essential oils in ever more varied contexts.

### 5.3. Choice of Encapsulation Materials

The choice of encapsulation materials is a critical aspect of nanotechnology system design. Biodegradable polymers such as PLGA (polylactide-co-glycolide) are frequently used due to their compatibility with medical and cosmetic applications (Almeida et al., 2019). Lipids, such as phospholipids, are also popular candidates due to their similarity to cell membranes.

Encapsulation materials need to be chosen according to desired properties, such as permeability, solubility and stability. It is also important to consider the biocompatibility of materials, particularly in the context of medical applications.

## 5.4. Factors influencing effectiveness

The effectiveness of nanotechnological encapsulation of essential oils is influenced by various factors  (Majeed et al., 2015).particle size is crucial, as it affects stability, bioavailability and release of active compounds (Yun et al., 2021). Smaller particles enable better tissue penetration, which is particularly important in medical and cosmetic applications.

The surface charge of nanoparticles can also influence their interaction with the biological environment. A modified surface charge can improve colloidal stability and particle distribution in formulations (Merino et al., 2019).

In addition, the manufacturing method used, as well as storage conditions, can affect encapsulation quality. Interactions between encapsulation materials and essential oils must be carefully studied to ensure effective, long-lasting encapsulation.

In summary, nanotechnological encapsulation techniques offer a promising approach to overcoming the challenges associated with the vulnerability of essential oils. Careful selection of materials and manufacturing methods, as well as an understanding of the factors influencing efficacy, are essential to fully exploit the potential of this technology in preserving and optimizing the properties of essential oils at the nanoscale. The next section of this book will explore in depth the specific applications of this innovative approach in the cosmetics industry.

## 6. Applications in the Cosmetics Industry
### 6.1. Improved skin penetration

Nanotechnological encapsulation of essential oils has revolutionized the cosmetics industry, significantly improving the cutaneous penetration of these precious compounds (Yun et al., 2021). Nanocapsules and nanoparticles enable the controlled release of essential oils, promoting deeper absorption into the skin's layers. This enhanced skin penetration capacity offers considerable advantages in delivering active ingredients directly where they are needed (Carvalho et al., 2016).

Nanotechnological formulations also make it possible to bypass traditional skin barriers, such as the stratum corneum, by facilitating the diffusion of essential oils across cell membranes.

This paves the way for more effective cosmetic products by maximizing the bioavailability of active compounds (Cimino et al., 2021).

## 6.2. Anti-Aging properties

In the cosmetics industry, nanotechnological encapsulation of essential oils offers sought-after anti-aging properties (Salem et al., 2022). The bioactive compounds present in certain essential oils, such as antioxidants, help combat the signs of skin aging.

Nanocapsules effectively protect these compounds from oxidation and interactions with other cosmetic ingredients. Encapsulated in this way, anti-aging essential oils can be integrated into specific formulations to target wrinkles, loss of elasticity and free radical damage (Lohani & Verma, 2022).

In short, nanotechnological encapsulation of essential oils has profoundly transformed the cosmetics industry, offering innovative solutions to combat the signs of ageing and improve product quality. The benefits of improved skin penetration and anti-aging properties testify to the revolutionary potential of this technology in the field of beauty and well-being. The following section will explore applications in the food industry, demonstrating the wide-ranging benefits of nanotechnological encapsulation of essential oils.

## 7. Food industry applications
### 7.1. Flavor stabilization

Nanotechnological encapsulation of essential oils is finding promising applications in the food industry, particularly for flavor stabilization (Gupta et al., 2016). Essential oils, rich in volatile aromatic compounds, can be sensitive to environmental conditions, leading to degradation of their characteristic aromas.

By encapsulating these essential oils on a nanometric scale, it is possible to protect the aromatic compounds from external factors such as oxygen, light and humidity. This encapsulation preserves the integrity of the aromas, guaranteeing long-term stability and controlled release when used in food products. (H. M. C. Marques, 2010).

### 7.2. Improved bioavailability

In the food industry, nanotechnological encapsulation of essential oils also offers advantages in terms of improved bioavailability (Delshadi et al., 2020). Nanoparticles and nanocapsules facilitate the homogeneous dispersion of essential oils in food matrices, ensuring uniform distribution of active compounds.

This improvement in bioavailability translates into more efficient absorption of essential oils by the body during food consumption. For example, in baked goods or beverages, nanotechnological encapsulation can promote the gradual release of flavors and bioactive compounds during digestion, maximizing their beneficial effects (Ozdal et al., 2020).

### 7.3. Examples of encapsulated foods

Nanotechnological encapsulation of essential oils has opened up new possibilities in the creation of innovative, functional foods. Notable examples include encapsulated flavor capsules in confections such as candy and chewing gum (Santos et al., 2014). These capsules slowly release flavors during chewing, offering an enhanced sensory experience (Santos et al., 2014).

Dairy products, such as yoghurts and desserts, can also benefit from nanotechnological encapsulation to improve flavor stability and ensure gradual release throughout consumption (Bylaitë et al., 2001). In addition, flavored drinks, condiments and snacks can incorporate this technology to offer longer-lasting flavor profiles and a better taste experience.

The use of encapsulated essential oils in foods not only preserves the freshness of flavors, but also opens up new dimensions in culinary creation. Consumers can enjoy tastier, more innovative food products, while benefiting from the functional properties of essential oils (Mishra et al., 2020).

In conclusion, nanotechnological encapsulation of essential oils has transformed the food industry, offering innovative solutions for stabilizing flavors and enhancing the bioavailability of active compounds. Examples of encapsulated foods demonstrate the potential of this technology to create more attractive, sensorially rich and functionally enhanced food products. The following section will examine the ethical and environmental implications of using nanotechnology in the encapsulation of essential oils.

## 8. Applications in Medicine

### 8.1. Drug Delivery

Nanotechnological encapsulation of essential oils has opened up new perspectives in drug delivery (Gagliardi et al., 2021).. Nanocapsules and nanoparticles offer an effective means of transporting therapeutic compounds to specific sites in the body. This approach, often referred to as "drug delivery", enables controlled, targeted release of active substances (Gagliardi et al., 2021).

Because of their antimicrobial, anti-inflammatory and antioxidant properties, essential oils can be encapsulated for use in the treatment of a variety of conditions. For example, encapsulated essential oils can be applied in the treatment of skin diseases, local inflammation or even anti-cancer applications (Cimino et al., 2021).

### 8.2. Targeted treatment

Nanotechnological encapsulation of essential oils offers the possibility of specifically targeting affected areas of the body. This is particularly important in cancer treatment, where precise drug delivery is crucial to minimize side effects on healthy tissue (Ojha & Mishra, 2023).

Nanoparticles can be designed to have a particular affinity for cancer cells, facilitating the selective delivery of anti-cancer essential oils. This approach improves treatment efficacy while reducing collateral damage to healthy tissue.

### 8.3. Applications in Alternative Medicine

The use of encapsulated essential oils is also finding its way into the field of alternative medicine, where natural approaches are increasingly in demand. The therapeutic properties of essential oils, such as their calming, analgesic and anti-inflammatory effects, can be exploited more precisely thanks to nanotechnological encapsulation (Mehta & MacGillivray, 2022).

For example, in medical aromatherapy, encapsulated essential oils can be used to treat specific disorders such as stress, anxiety or sleep disorders. The controlled release of active compounds enables prolonged diffusion of their beneficial effects, offering alternative solutions for improving mental and physical well-being.

In summary, nanotechnological encapsulation of essential oils has important applications in the medical field. From drug delivery to alternative medicine, this approach offers innovative solutions for improving treatment efficacy while minimizing adverse effects. The next section will explore the ethical considerations and potential challenges associated with the use of nanotechnology in the medical field.

## 9. Environmental impact

### 9.1. Ecotoxicity of Nanomaterials

The increasing use of nanotechnology raises concerns about the potential ecotoxicity of nanomaterials, including those used in the encapsulation of essential oils (Karabasz et al., 2019; Tian et al., 2017). Nanoparticles can pose risks to the environment, particularly when released into wastewater or soil.

Studies are underway to assess the effects of nanomaterials on aquatic and terrestrial ecosystems (Lopes et al., 2017). It is crucial to understand how these nanoparticles interact with living organisms, whether microorganisms, plants or animals. Environmental monitoring and ongoing research are essential to minimize potential negative impacts and ensure responsible use of nanotechnology (Figueiredo et al., 2019).

### 9.2. Sustainable Approaches to Nanotechnology

Faced with environmental concerns, researchers and industries are exploring sustainable approaches to nanotechnology. This includes the development of biodegradable and environmentally-friendly nanomaterials. Efforts are also being made to design cleaner, energy-efficient manufacturing processes, thereby reducing nanotechnology's overall environmental footprint (Jafari et al., 2017).

Designing nanomaterials that minimize potential risks while maximizing their benefits is a priority. In-depth research into the sustainability of nanomaterials and associated processes is helping to steer industry towards more environmentally responsible practices.

### 9.3. Ethical considerations

Advances in nanotechnology also raise important ethical considerations. The use of nanomaterials, including in the encapsulation of essential oils, requires reflection on the ethical implications of this technology.

It is essential to guarantee the safety of consumers, industrial workers and the environment in the development and use of nanotechnology products. This calls for greater transparency, rigorous risk assessment and appropriate regulations to govern this emerging technology.

In addition, it is important to take account of social justice issues, ensuring that the benefits of nanotechnology are distributed fairly and that its use does not exacerbate existing inequalities.

In conclusion, the environmental impacts of nanotechnology, including in the field of essential oil encapsulation, require careful management. Sustainable approaches and ethical considerations are essential to ensure that the benefits of nanotechnology do not compromise the health of the planet or create social injustices. The next section will explore future prospects for nanotechnological encapsulation of essential oils, highlighting the challenges and opportunities ahead.

## 10. Future prospects

### 10.1.    Research in progress

Ongoing research into the nanotechnological encapsulation of essential oils is exploring many exciting avenues. Scientists are striving to improve encapsulation techniques, focusing on the precision and stability of nanoparticles. In-depth studies are also examining the interactions between nanomaterials and the active compounds in essential oils, aiming to optimize controlled release for specific applications.

Advances in nanomaterial characterization and understanding of release mechanisms are helping to refine formulations, paving the way for more precise and effective applications.

## 10.2.    Expected innovations

Future innovations in nanotechnological encapsulation of essential oils should include significant improvements in nanomaterial design and manufacturing methods. The possibility of creating smaller, more stable nanocapsules offering even more controlled release will open up new application prospects.

Innovations in the choice of encapsulation materials could also lead to more sustainable, biodegradable and environmentally-friendly solutions. The integration of emerging technologies, such as artificial intelligence and computer modeling, could accelerate the process of designing and optimizing formulations.

Exploring new synergies between nanotechnology and other fields, such as synthetic biology or regenerative medicine, could also lead to innovative applications, transforming the way we use essential oils at the nanoscale.

## 10.3.    Growth Potential

The growth potential of nanotechnological encapsulation of essential oils is immense. With growing demand for more effective, sustainable and environmentally-friendly products, this technology offers solutions for many sectors, including cosmetics, food and medicine.

Future growth could also be fueled by a better understanding of the therapeutic benefits of essential oils at the nanoscale, paving the way for more specific and personalized medical applications.

Collaboration between academic research, industry and regulators will be crucial to fully exploiting the growth potential of this technology, while guaranteeing the safety and sustainability of its use.

In conclusion, the future prospects for nanotechnological encapsulation of essential oils are full of promise. Ongoing research, expected innovations and growth potential are paving the way for a new era of applications and benefits in various fields, making this technology a key driver of innovation in the years to come.

# 11. Conclusion

## 11.1.　　Summary of key points

This in-depth exploration of nanotechnological encapsulation of essential oils has highlighted several key aspects of this innovative technology. We began with an in-depth introduction, highlighting the importance of essential oils and nanotechnology in various fields.

By exploring the fundamentals of essential oils, we have defined their origin, extraction, therapeutic properties and current applications. Understanding these fundamentals has laid the foundations for understanding the challenges associated with their vulnerability to environmental conditions.

We then examined the encapsulation needs of essential oils, highlighting the vulnerability of these compounds and the crucial benefits of encapsulation, including improved stability and shelf life.

The section on nanotechnological encapsulation techniques detailed the different approaches, from nanoparticles to manufacturing techniques, choice of encapsulation materials and factors influencing efficiency.

Specific applications in the cosmetics and food industries have illustrated the transformative potential of nanotechnological encapsulation of essential oils, whether to enhance skin penetration, offer anti-aging properties or stabilize food flavors.

We then turned to the environmental implications, highlighting the challenges of nanomaterial ecotoxicity and emphasizing sustainable approaches to nanotechnology. Ethical considerations were also highlighted, underscoring the need for responsible use of this technology.

Finally, future prospects highlighted ongoing research, expected innovations and the strong growth potential of nanotechnological encapsulation of essential oils in the years to come.

## 11.2.　　Outlook for the future

Nanotechnological encapsulation of essential oils represents an exciting frontier of science and technology. As we continue to explore new avenues, in-depth research and multidisciplinary collaborations will be crucial to maximize the benefits of this technology while mitigating potential risks.

Future innovations could transform the way we formulate cosmetics, develop food products and deliver medicines. Advances in the sustainability of nanomaterials and a detailed understanding of interactions with the active compounds in essential oils will pave the way for even more precise and effective applications.

It is essential to remain attentive to ethical and environmental implications, ensuring that this technology makes a positive contribution to society while minimizing potential risks.

In conclusion, nanotechnological encapsulation of essential oils offers considerable potential to significantly improve our daily lives. As this technology evolves, it will continue to shape the future of the cosmetics, food and medical industries, bringing innovative solutions and sustainable benefits. The road ahead looks exciting, and ongoing research in this field promises to open up exciting new perspectives and opportunities.

**References**

Ali, B., Al-Wabel, N. A., Shams, S., Ahamad, A., Khan, S. A., & Anwar, F. (2015). Essential oils used in aromatherapy: A systemic review. *Asian Pacific Journal of Tropical Biomedicine, 5*(8), 601-611.

Almeida, K. B., Ramos, A. S., Nunes, J. B. B., Silva, B. O., Ferraz, E. R. A., Fernandes, A. S., Felzenszwalb, I., Amaral, A. C. F., Roullin, V. G., & Falcão, D. Q. (2019). PLGA nanoparticles optimized by Box-Behnken for efficient encapsulation of therapeutic Cymbopogon citratus essential oil. *Colloids and Surfaces B: Biointerfaces, 181*, 935-942. https://doi.org/10.1016/j.colsurfb.2019.06.010

Assadpour, E., & Mahdi Jafari, S. (2019). A systematic review on nanoencapsulation of food bioactive ingredients and nutraceuticals by various nanocarriers. *Critical Reviews in Food Science and Nutrition, 59*(19), 3129-3151. https://doi.org/10.1080/10408398.2018.1484687

Aydeniz-Guneser, B. (2020). Cold pressed orange (Citrus sinensis) oil. In *Cold Pressed Oils* (p. 129-146). Elsevier.

Aziz, Z. A., Ahmad, A., Setapar, S. H. M., Karakucuk, A., Azim, M. M., Lokhat, D., Rafatullah, M., Ganash, M., Kamal, M. A., & Ashraf, G. M. (2018). Essential oils: Extraction techniques, pharmaceutical and therapeutic potential-a review. *Current drug metabolism, 19*(13), 1100-1110.

Barradas, T. N., & de Holanda e Silva, K. G. (2021). Nanoemulsions of essential oils to improve solubility, stability and permeability: A review. *Environmental Chemistry Letters, 19*(2), 1153-1171. https://doi.org/10.1007/s10311-020-01142-2

Bylaitë, E., Rimantas Venskutonis, P., & Maþdþierienë, R. (2001). Properties of caraway (Carum carvi L.) essential oil encapsulated into milk protein-based matrices. *European Food Research and Technology, 212*(6), 661-670. https://doi.org/10.1007/s002170100297

Carvalho, I. T., Estevinho, B. N., & Santos, L. (2016). Application of microencapsulated essential oils in cosmetic and personal healthcare products - a review. *International Journal of Cosmetic Science, 38*(2), 109-119. https://doi.org/10.1111/ics.12232

Cavalcanti, A., & Freitas, R. A. (2005). Nanorobotics control design: A collective behavior approach for medicine. *IEEE Transactions on Nanobioscience, 4*(2), 133-140.

Charles, D. J., & Simon, J. E. (1990). Comparison of extraction methods for the rapid determination of essential oil content and composition of basil. *Journal of the American Society for Horticultural Science, 115*(3), 458-462.

Chemat, F., & Sawamura, M. (2010). Techniques for oil extraction. *Citrus essential oils: Flavor and fragrance*, 9-36.

Cimino, C., Maurel, O. M., Musumeci, T., Bonaccorso, A., Drago, F., Souto, E. M. B., Pignatello, R., & Carbone, C. (2021). Essential Oils: Pharmaceutical Applications and Encapsulation Strategies into Lipid-Based Delivery Systems. *Pharmaceutics, 13*(3), Article 3. https://doi.org/10.3390/pharmaceutics13030327

Damian, P., & Damian, K. (1995). *Aromatherapy: Scent and psyche: Using essential oils for physical and emotional well-being*. Inner Traditions/Bear & Co.

De Oliveira, M. S., Silva, S. G., da Cruz, J. N., Ortiz, E., da Costa, W. A., Bezerra, F. W. F., Cunha, V., Cordeiro, R., de Jesus Chaves Neto, A., & de Aguiar Andrade, E. (2019). Supercritical CO2 application in essential oil extraction. *Industrial Applications of Green Solvents, 2*, 1-28.

Delshadi, R., Bahrami, A., Tafti, A. G., Barba, F. J., & Williams, L. L. (2020). Micro and nano-encapsulation of vegetable and essential oils to develop functional food products with improved nutritional profiles. *Trends in Food Science & Technology*, *104*, 72-83. https://doi.org/10.1016/j.tifs.2020.07.004

Demirer, G. S., Silva, T. N., Jackson, C. T., Thomas, J. B., W. Ehrhardt, D., Rhee, S. Y., Mortimer, J. C., & Landry, M. P. (2021). Nanotechnology to advance CRISPR-Cas genetic engineering of plants. *Nature Nanotechnology, 16*(3), 243-250.

Deng, J., Lu, X., Liu, L., Zhang, L., & Schmidt, O. G. (2016). Introducing rolled-up nanotechnology for advanced energy storage devices. *Advanced Energy Materials, 6*(23), 1600797.

Djilani, A., & Dicko, A. (2012). The therapeutic benefits of essential oils. *Nutrition, well-being and health, 7*, 155-179.

Dupuy, J.-P., & Roure, F. (2004). Les nanotechnologies: Éthique et prospective industrielle. *Conseil Général des Mines and Conseil Général des Technologies de l'Information.*

El Asbahani, A., Miladi, K., Badri, W., Sala, M., Addi, E. A., Casabianca, H., El Mousadik, A., Hartmann, D., Jilale, A., & Renaud, F. (2015). Essential oils: From extraction to encapsulation. *International journal of pharmaceutics, 483*(1-2), 220-243.

El Mostain, A. (2022). Et l'alambic créa l'eau de vie. Retour sur l'histoire technique de l'alambic. *e-Phaïstos. Revue d'histoire des techniques/Journal of the history of technology, X-1.*

Fangmeier, M., Lehn, D., Jachetti Maciel, M., & Souza, C. (2019). Encapsulation of Bioactive Ingredients by Extrusion with Vibrating Technology: Advantages and Challenges. *Food and Bioprocess Technology, 12*, 1-15. https://doi.org/10.1007/s11947-019-02326-7

Figueiredo, J., Oliveira, T., Ferreira, V., Sushkova, A., Silva, S., Carneiro, D., Cardoso, D. N., Gonçalves, S. F., Maia, F., Rocha, C., Tedim, J., Loureiro, S., & Martins, R. (2019). Toxicity of innovative antifouling nano-based solutions to marine species. *Environmental Science: Nano*, *6*(5), 1418-1429. https://doi.org/10.1039/C9EN00011A

Gagliardi, A., Giuliano, E., Venkateswararao, E., Fresta, M., Bulotta, S., Awasthi, V., & Cosco, D. (2021). Biodegradable Polymeric Nanoparticles for Drug Delivery to Solid Tumors. *Frontiers in Pharmacology*, *12*, 601626. https://doi.org/10.3389/fphar.2021.601626

Gonzalez-Burgos, E., & Gomez-Serranillos, M. (2012). Terpene compounds in nature: A review of their potential antioxidant activity. *Current medicinal chemistry*, *19*(31), 5319-5341.

Gupta, S., Khan, S., Muzafar, M., Kushwaha, M., Yadav, A. K., & Gupta, A. P. (2016). 6 - Encapsulation: Entrapping essential oil/flavors/aromas in food. In A. M. Grumezescu (Ed.), *Encapsulations* (pp. 229-268). Academic Press. https://doi.org/10.1016/B978-0-12-804307-3.00006-5

Guzmán, E., & Lucia, A. (2021). Essential Oils and Their Individual Components in Cosmetic Products. *Cosmetics*, *8*(4), Article 4. https://doi.org/10.3390/cosmetics8040114

Jafari, S. M., Katouzian, I., & Akhavan, S. (2017). 15-Safety and regulatory issues of nanocapsules. In S. M. Jafari (Ed.), *Nanoencapsulation Technologies for the Food and Nutraceutical Industries* (pp. 545-590). Academic Press. https://doi.org/10.1016/B978-0-12-809436-5.00015-X

James, G. (2023). *Introduction to nanotechnology*. Gilad James Mystery School.

Jugreet, B. S., Suroowan, S., Rengasamy, R. K., & Mahomoodally, M. F. (2020). Chemistry, bioactivities, mode of action and industrial applications of essential oils. *Trends in Food Science & Technology*, *101*, 89-105.

Karabasz, A., Szczepanowicz, K., Cierniak, A., Mezyk-Kopec, R., Dyduch, G., Szczęch, M., Bereta, J., & Bzowska, M. (2019). In vivo Studies on Pharmacokinetics, Toxicity and Immunogenicity of Polyelectrolyte Nanocapsules Functionalized with Two Different Polymers: Poly-L-Glutamic Acid or PEG. *International Journal of Nanomedicine, 14,* 9587-9602. https://doi.org/10.2147/IJN.S230865

Kayaci, F., & Uyar, T. (2011). Solid Inclusion Complexes of Vanillin with Cyclodextrins: Their Formation, Characterization, and High-Temperature Stability. *Journal of Agricultural and Food Chemistry, 59*(21), 11772-11778. https://doi.org/10.1021/jf202915c

Khandve, P. (2014). Nanotechnology for building material. *International Journal of Basic and Applied Research, 4,* 146-151.

Khanna, A. (2008). Nanotechnology in high performance paint coatings. *Asian J. Exp. Sci, 21*(2), 25-32.

Lardry, J.-M., & Haberkorn, V. (2007). Aromatherapy and essential oils. *Kinésithérapie, la revue, 7*(61), 14-17.

Laurain-Mattar, D., Couic-Marinier, F., & Aribi-Zouioueche, L. (2022). Huile essentielle d'Origan vulgaire. *Actualités Pharmaceutiques, 61*(619), 57-59.

Lohani, A., & Verma, A. (2022). 6 - Lipid vesicles: Potential nanocarriers for the delivery of essential oils to combat skin aging. In S. H. Mohd Setapar, A. Ahmad, & M. Jawaid (Eds.), *Nanotechnology for the Preparation of Cosmetics Using Plant-Based Extracts* (pp. 131-156). Elsevier. https://doi.org/10.1016/B978-0-12-822967-5.00006-0

Lopes, L. Q. S., Santos, C. G., de Almeida Vaucher, R., Raffin, R. P., da Silva, A. S., Baretta, D., Maccari, A. P., Giombelli, L. C. D. D., Volpato, A., Arruda, J., de Ávila Scheeren, C., Baldisserotto, B., & Santos, R. C. V. (2017). Ecotoxicology of Glycerol Monolaurate nanocapsules. *Ecotoxicology and Environmental Safety, 139,* 73-77. https://doi.org/10.1016/j.ecoenv.2017.01.019

Majeed, H., Bian, Y.-Y., Ali, B., Jamil, A., Majeed, U., Khan, Q. F., Iqbal, K. J., Shoemaker, C. F., & Fang, Z. (2015). Essential oil encapsulations: Uses, procedures, and trends. *Rsc Advances, 5*(72), 58449-58463.

Mamadalieva, N. Z., Akramov, D. K., Ovidi, E., Tiezzi, A., Nahar, L., Azimova, S. S., & Sarker, S. D. (2017). Aromatic medicinal plants of the Lamiaceae family from Uzbekistan: Ethnopharmacology, essential oils composition, and biological activities. *Medicines, 4*(1), 8.

Mansour, M. M., El-Hefny, M., Salem, M. Z., & Ali, H. M. (2020). The biofungicide activity of some plant essential oils for the cleaner production of model linen fibers similar to those used in ancient Egyptian mummification. *Processes, 8*(1), 79.

Marques, F. M., Figueira, M. M., Schmitt, E. F. P., Kondratyuk, T. P., Endringer, D. C., Scherer, R., & Fronza, M. (2019). In vitro anti-inflammatory activity of terpenes via suppression of superoxide and nitric oxide generation and the NF-κB signalling pathway. *Inflammopharmacology, 27*, 281-289.

Marques, H. M. C. (2010). A review on cyclodextrin encapsulation of essential oils and volatiles. *Flavour and Fragrance Journal, 25*(5), 313-326. https://doi.org/10.1002/ffj.2019

Maurya, A., Prasad, J., Das, S., & Dwivedy, A. K. (2021). Essential oils and their application in food safety. *Frontiers in Sustainable Food Systems, 5*, 653420.

Mehta, S., & MacGillivray, M. (2022). Aromatherapy in Textiles: A Systematic Review of Studies Examining Textiles as a Potential Carrier for the Therapeutic Effects of Essential Oils. *Textiles, 2*(1), Article 1. https://doi.org/10.3390/textiles2010003

Merino, N., Berdejo, D., Bento, R., Salman, H., Lanz, M., Maggi, F., Sánchez-Gómez, S., García-Gonzalo, D., & Pagán, R. (2019). Antimicrobial efficacy of Thymbra capitata (L.) Cav. Essential oil loaded in self-assembled zein nanoparticles in combination with heat. *Industrial Crops and Products, 133,* 98-104. https://doi.org/10.1016/j.indcrop.2019.03.003

Miguel, M. G., Lourenço, J. P., & Faleiro, M. L. (2020). Superparamagnetic Iron Oxide Nanoparticles and Essential Oils: A New Tool for Biological Applications. *International Journal of Molecular Sciences, 21*(18), Article 18. https://doi.org/10.3390/ijms21186633

Mishra, A. P., Devkota, H. P., Nigam, M., Adetunji, C. O., Srivastava, N., Saklani, S., Shukla, I., Azmi, L., Shariati, M. A., Melo Coutinho, H. D., & Mousavi Khaneghah, A. (2020). Combination of essential oils in dairy products: A review of their functions and potential benefits. *LWT, 133,* 110116. https://doi.org/10.1016/j.lwt.2020.110116

Mustafa, I. F., & Hussein, M. Z. (2020). Synthesis and Technology of Nanoemulsion-Based Pesticide Formulation. *Nanomaterials, 10*(8), Article 8. https://doi.org/10.3390/nano10081608

Ojha, S., & Mishra, S. (2023). Nanostructured Lipid Carriers for Targeting Central Nervous System: Recent Advancements. *Micro and Nanosystems, 15*(2), 82-91. https://doi.org/10.2174/1876402915666230518121949

Ortner, V., Kaspar, C., Halter, C., Töllner, L., Mykhaylyk, O., Walzer, J., Günzburg, W. H., Dangerfield, J. A., Hohenadl, C., & Czerny, T. (2012). Magnetic field-controlled gene expression in encapsulated cells. *Journal of Controlled Release, 158*(3), 424-432. https://doi.org/10.1016/j.jconrel.2011.12.006

Ozdal, T., Yolci-Omeroglu, P., & Tamer, E. C. (2020). 6-Role of Encapsulation in Functional Beverages. In A. M. Grumezescu & A. M. Holban (Eds.), *Biotechnological Progress and Beverage Consumption* (pp. 195-232). Academic Press. https://doi.org/10.1016/B978-0-12-816678-9.00006-0

Pandey, P. (2022). Role of nanotechnology in electronics: A review of recent developments and patents. *Recent Patents on Nanotechnology, 16*(1), 45-66.

Pandit, J., Aqil, Mohd., & Sultana, Y. (2016). 14-Nanoencapsulation technology to control release and enhance bioactivity of essential oils. In A. M. Grumezescu (Ed.), *Encapsulations* (pp. 597-640). Academic Press. https://doi.org/10.1016/B978-0-12-804307-3.00014-4

Perinelli, D. R., Cespi, M., & Bonacucina, G. (2019). 19-Nanostructures of chemical biodegradable polymers and their derivatives for encapsulation of food ingredients. In S. M. Jafari (Ed.), *Biopolymer Nanostructures for Food Encapsulation Purposes* (pp. 581-606). Academic Press. https://doi.org/10.1016/B978-0-12-815663-6.00019-7

Ríos, J.-L. (2016). Chapter 1 - Essential Oils: What They Are and How the Terms Are Used and Defined. In V. R. Preedy (Ed.), *Essential Oils in Food Preservation, Flavor and Safety* (pp. 3-10). Academic Press. https://doi.org/10.1016/B978-0-12-416641-7.00001-8

Salem, M. A., Manaa, E. G., Osama, N., Aborehab, N. M., Ragab, M. F., Haggag, Y. A., Ibrahim, M. T., & Hamdan, D. I. (2022). Coriander (Coriandrum sativum L.) essential oil and oil-loaded nano-formulations as an anti-aging potentiality via TGFβ/SMAD pathway. *Scientific Reports*, *12*(1), Article 1. https://doi.org/10.1038/s41598-022-10494-4

Santos, M. G., Carpinteiro, D. A., Thomazini, M., Rocha-Selmi, G. A., da Cruz, A. G., Rodrigues, C. E. C., & Favaro-Trindade, C. S. (2014). Coencapsulation of xylitol and menthol by double emulsion followed by complex coacervation and microcapsule application in chewing gum. *Food Research International*, *66*, 454-462. https://doi.org/10.1016/j.foodres.2014.10.010

Sarkic, A., & Stappen, I. (2018). Essential oils and their single compounds in cosmetics-A critical review. *Cosmetics*, *5*(1), 11.

Scimeca, D. (2006). *Les plantes du bonheur: Le coup de pouce des plantes contre tous les coups de blues*. Alpen Editions sam.

Shoukat, R., & Khan, M. I. (2021). Carbon nanotubes: A review on properties, synthesis methods and applications in micro and nanotechnology. *Microsystem Technologies*, 1-10.

Silva, G. A. (2004). Introduction to nanotechnology and its applications to medicine. *Surgical neurology*, *61*(3), 216-220.

Silva-Flores, P. G., Galindo-Rodríguez, S. A., Pérez-López, L. A., & Álvarez-Román, R. (2023). Development of Essential Oil-Loaded Polymeric Nanocapsules as Skin Delivery Systems: Biophysical Parameters and Dermatokinetics Ex Vivo Evaluation. *Molecules*, *28*(20), Article 20. https://doi.org/10.3390/molecules28207142

Sinha, R., Kim, G. J., Nie, S., & Shin, D. M. (2006). Nanotechnology in cancer therapeutics: Bioconjugated nanoparticles for drug delivery. *Molecular cancer therapeutics*, *5*(8), 1909-1917.

Siva, S., Li, C., Cui, H., Meenatchi, V., & Lin, L. (2020). Encapsulation of essential oil components with methyl-β-cyclodextrin using ultrasonication: Solubility, characterization, DPPH and antibacterial assay. *Ultrasonics Sonochemistry*, *64*, 104997. https://doi.org/10.1016/j.ultsonch.2020.104997

Suffredini, G., East, J., & Levy, L. (2013). New Applications of Nanotechnology for Neuroimaging. *AJNR. American journal of neuroradiology*, *35*. https://doi.org/10.3174/ajnr.A3543

Tian, B., Wang, Q., Su, Q., Feng, W., & Li, F. (2017). In vivo biodistribution and toxicity assessment of triplet-triplet annihilation-based upconversion nanocapsules. *Biomaterials*, *112*, 10-19. https://doi.org/10.1016/j.biomaterials.2016.10.008

Timilsena, Y. P., Akanbi, T. O., Khalid, N., Adhikari, B., & Barrow, C. J. (2019). Complex coacervation: Principles, mechanisms and applications in microencapsulation. *International Journal of Biological Macromolecules*, *121*, 1276-1286. https://doi.org/10.1016/j.ijbiomac.2018.10.144

Torres, C. S. (2003). *Alternative lithography: Unleashing the potentials of nanotechnology*. Springer Science & Business Media.

Turek, C., & Stintzing, F. C. (2013). Stability of essential oils: A review. *Comprehensive reviews in food science and food safety, 12*(1), 40-53.

Venugopal, J., Prabhakaran, M. P., Low, S., Choon, A. T., Zhang, Y., Deepika, G., & Ramakrishna, S. (2008). Nanotechnology for nanomedicine and delivery of drugs. *Current pharmaceutical design, 14*(22), 2184-2200.

Wen, P., Zong, M.-H., Linhardt, R. J., Feng, K., & Wu, H. (2017). Electrospinning: A novel nano-encapsulation approach for bioactive compounds. *Trends in Food Science & Technology, 70*, 56-68. https://doi.org/10.1016/j.tifs.2017.10.009

Yuan, C., Wang, Y., Liu, Y., & Cui, B. (2019). Physicochemical characterization and antibacterial activity assessment of lavender essential oil encapsulated in hydroxypropyl-beta-cyclodextrin. *Industrial Crops and Products, 130*, 104-110. https://doi.org/10.1016/j.indcrop.2018.12.067

Yun, P., Devahastin, S., & Chiewchan, N. (2021). Microstructures of encapsulates and their relations with encapsulation efficiency and controlled release of bioactive constituents: A review. *Comprehensive Reviews in Food Science and Food Safety, 20*(2), 1768-1799. https://doi.org/10.1111/1541-4337.12701

Zerrougui, K., & Amira, A. (2023). *Biosynthesis of metallic nanoparticles based on mint leaf essential oil*. UNIVERSITE KASDI MERBAH OUARGLA.

# yes I want morebooks!

Buy your books fast and straightforward online - at one of world's fastest growing online book stores! Environmentally sound due to Print-on-Demand technologies.

Buy your books online at
## www.morebooks.shop

Kaufen Sie Ihre Bücher schnell und unkompliziert online – auf einer der am schnellsten wachsenden Buchhandelsplattformen weltweit! Dank Print-On-Demand umwelt- und ressourcenschonend produzi ert.

Bücher schneller online kaufen
## www.morebooks.shop

info@omniscriptum.com
www.omniscriptum.com

Printed by Books on Demand GmbH, Norderstedt / Germany